How To Delete Books Off Your Kindle

A Simple Step-By-Step Guide On How To Delete Books From Your Kindle

I0504758

Table Of Contents

Introduction

In this book you will find all the information you need when it comes to deleting and archiving books off of your Kindle. In it are step-by-step guides to help ensure that the process is easy to follow and foolproof, even if you're not particularly used to technology.

Deleting books off of Kindle isn't the most straightforward process, but with the guide we have provided in this book? It shouldn't take you more than 5 to 10 minutes to accomplish. No need to phone a customer service representative or get stressed over it!

Let's begin the journey.

DELETE BOOKS OFF OF KINDLE EASILY

Kindle is one of the most popular e-readers currently available. It is used widely and by people of all ages. The convenience and accessibility it offers when it comes to digital literature is unparalleled—you literally have hundreds of thousands of books available right at your fingertips. Not to mention the fact that users get to enjoy numerous samples and even free books as long as they have an account with Amazon.

Needless to say, it's quite easy to hoard digital books on one's e-reader. Purchasing it is comparably more affordable compared to hard copies and much like real books, you also have the option of borrowing and lending within your social circle. It's also a must for people who are always on the go and want to avoid having to bring a heavy book with them. After all, the e-reader itself is slim and sleek—easily fitting into purses and bags to be taken along.

The same applies to people who want to read and keep a lot of books, but don't have enough space in their homes to accommodate it. Perhaps you're a university student living in a shared dorm space, where the only available spots for books available are those you need for classes. Where will you keep your trilogies, your favorite series, your unread pile, and your new book pile?

Quite the nightmare, isn't it?

But you need not worry about all that if you have a Kindle. You just need one device and it's more than enough to hold all of the above and so much more. The best bit? You can take all of your books with you wherever you go, allowing you to read whenever your busy schedule allows. On the commute, during breaks, and even in between waiting for classes.

Besides, it isn't just leisure reads that you can add to your reader. If you need plenty of references for your schoolwork and classes, you'll find plenty of those on Amazon available for download. These can be easily added to your reader, allowing you to study different subjects without having to weigh yourself down with one too many books.

Now, if you're quite the voracious reader, much like myself, then your Kindle is likely to contain plenty of new downloads, unread books, and samples that you've been eyeing. Whilst there's nothing wrong with having a lot of options, it can become overwhelming after some time.

Think of this as akin to having piles upon piles of unread books on your desk or gathering dust in your bookshelf, wondering when you'll have enough time to go through them. It's no different from buying and hoarding. I'm sure you have tried to do a clean out of your collection—but ended up getting new purchases in the process.

How long has it been since you thoroughly checked your account and gave it a nice sweep? A few months? A few years? Since we all lead such busy lives, it is likely that we don't pay enough attention to our e-readers other than to download new things or try and continue what we have been reading beforehand.

No shame in that, but it's about time for a clean out. An organized Kindle means you can see what you still have left to read and lessen the stress brought on by the chaos of titles that you're no longer sure if you're still keen on reading. Now, "purging" would have been a simple task of select and delete had Amazon made the process that simple and intuitive.

Sadly, it is not.

Again, if you've ever tried doing this then you know the process itself can be quite confusing (and tedious!) especially if you're a new user and this is your first time doing it. Fret not, however, in this book we will be compiling a few different ways you can do this to help make sure that you accomplish the task with very little difficulty.

Deleting books off your kindle doesn't mean you'll have to say goodbye to them forever—you can still re-download every title, but deleting them means you won't see them on your e-reader anymore. It's no different from archiving and any time you want to read those old books, you would be able to easily recover them.

Lastly, I shall also be providing you with tips on how you can keep your Kindle as organized as possible. This should allow you to enjoy your purchases better!

METHOD 1: CLEARING ITEMS FROM YOUR LIBRARY

This method is for deleting titles off of your Kindle permanently. Note that by doing this, you will have to pay and download the books again if you want to re-read them. So make sure you're certain of what you're doing before continuing with this process.

1. Start by logging into your content library through the Amazon website. From here, access "manage your content and devices". Double check the page to make sure you're on the right one.

2. On the "Your Content" page, you should be able to see an inventory of all the items in your library. Select the different books you want to delete by ticking the empty boxes located on the far left of the screen. You can click on the box a second time should you choose the wrong book.

3. Don't worry! This is only the selection phase of the process, clicking a book accidentally will not automatically delete it. To hasten the process for yourself, create a shortlist of read books before you even login. This helps cut down on both the choosing time and possibility of you selecting a title you haven't read just yet.

4. Once you're happy with your selections and have finished double checking everything, you can proceed to deleting them. To do this, find the orange button for "DELIVER" and another orange button labeled "DELETE". Both are located at the top left side of your screen and can be a bit confusing so make sure you check before clicking.

5. Click the delete button to remove the selected items. You will receive a prompt asking you to confirm the action. Click on "Yes, delete permanently" so ensure proper deletion. Note that this final prompt allows you to go back if you happen to change your mind, so click cancel if you feel like you want to add or remove titles from your selection.

6. Finally, another pop up will confirm your action once you've deleted everything. Simply click on "OK" to remove it from your screen.

SYNCING DEVICES

Now that you've finished removing the books off of your Kindle account, you would want the changes to reflect in all of your devices that has it. To do this in a convenient manner, you simply need to sync your device. Here's how:

- Get your device and click on the Kindle application.

- Once open, click on the three dots that will appear on the upper right hand of your screen.

- Clicking on these will open a menu and give you the option to "Sync and Check for Items". Click on this option the books you have removed will be deleted from your device as well.

METHOD 2: DELETING BOOKS THROUGH YOUR DEVICE

If you currently don't have access to a desktop, you can still proceed with clearing out your Kindle library through the use of your device. Here are the steps you should follow:

1. Access the app by tapping on its icon. Once the Home Screen opens, you should be able to see and access your Kindle library like usual. Search for the title you wish to delete by scrolling on the screen up and down.

2. You can also search for the title by clicking on the magnifying glass. This will open a bar where you can type the book's title.

3. After you've located the book you plan on deleting, press your finger on it and hold for a few seconds until another menu appears on your screen. This menu should provide you with different options, including "Remove from device."

4. Select the remove option by lightly tapping on it. Note that by doing this, you will effectively remove the book from your device but it will still be available in your cloud. This means you'll be able to re-download it again if need be.

Compared to the previous option, removing books using your Kindle device is much less permanent. So if you're prone to indecision, a feeling I can certainly relate with, use this option instead. This should help you avoid accidental permanent deletions or needing to re-purchase a book that you deleted.

With that in mind, is there a way for you to archive books instead of deleting them?

METHOD 3: ARCHIVING BOOKS ON YOUR KINDLE

Want to organize your Kindle library and only display books you're currently or will be reading? Well, if you're not keen on deleting older titles or book samples, the next best option will be to archive them instead. As the name implies, this means you'll simply be "stowing" these titles away where they aren't readily displayed each time you access your account.

This is great for folks who do become quite attached to every title they purchase and feel as if deleting is a waste of the money they used to pay for it. Besides, who doesn't enjoy re-reading certain books we enjoyed once?

To archive book titles on your Kindle, here are the steps you need to follow:

1. Start by powering on your Kindle. Click on the home icon in order to access your home screen. On your home screen, you should be able to see all of the books in your library.

2. Go through the different titles. If your library contains a lot, it is highly suggested that you make a list of all the titles you want to archive and those you're ready to delete permanently.

3. Scroll through the different titles. On a piece of paper, create a column for books you're archiving and another one for those you plan on deleting. Write down titles as you go until you've managed to go through your entire collection.

4. It pays to double check so make sure you do before proceeding with the archiving process.

5. Once you're sure of your list, click on the first title on your list. Make sure it has been properly selected by checking for an underline—this is a thick black line that appears after you click on a book.

6. Next, click on the "right" direction or arrow on your Kindle's 5-way controller. Pressing this would open up a menu with different options for you to choose from. To archive, click on "remove from device".

7. Doing this removes it from your device, but archives the book in your account. The record of your purchase and download will be maintained as well.

8. If you wish to restore the archived book, all you need to do is go to the "Archived Items" category of your Kindle screen. Scroll through the different titles until you find the one you're looking for then click on the center button of your controller. This should start the re-download of your book.

METHOD 4: REMOVING YOUR KINDLE REGISTRATION

This particular method is meant for purposes such as completely clearing out your device if you are switching from one e-reader to another, selling your Kindle, if your device gets lost, or if it ends up stolen. Removing your Kindle registration removes all of the item in your device.

Note that doing this means you will no longer be able to access the content and they will be removed from your cloud as well. The only way to gain access to the titles you once had saved would be to re-register your Kindle. Keep that in mind before going through this process.

The same goes for any personal information you have saved in your e-reader thus making it a necessary step if the device is no longer in your hands. For that, here are the steps you need to follow:

1. Since you no longer have the device in your hands, you will need to open your account on a desktop computer. Proceed to the "Manage Your Content and Devices" option. You will need to be logged in to your account in order to do this so make sure you have your details memorized or saved somewhere.

2. Next, select "Your Devices". Clicking this should bring you to a page that will list all of your devices that are registered to your account. Select the particular device you're deregistering by clicking or tapping on it once. Note that clicking once will not immediately finish the process so if you happen to click the wrong one accidently, don't fret.

3. The selected device will have an orange outline and pink background so you can be sure that you have chosen the right one. The unselected ones will remain the same.

4. Once you've chosen the devices, click on the deregister button. You will find this just below the list of your devices,

on the left side of your screen. Again, double check the device you've chosen before clicking on this button.

5. After you tap on this button, your Kindle will then be disconnected from your Amazon account. Any books, whether read or unread, that you have purchased will be automatically removed from your Kindle as well. Do understand that deregistering your reader also means that you will not be able to download and purchase any new content unless you register a new account with Amazon.

If you're only planning on removing a few titles off of your Kindle, this particular method might be too drastic for that. It is only recommended that you use this if you: intend on starting over and want a clean slate for your collection or if your device gets stolen and you want to make sure that every content has been removed from it.

HOW TO RE-REGISTER YOUR KINDLE

Should you need to register a different account for your Kindle e-reader, the process is quite similar to that of deregistering. Here's what you need to do:

1. Open your Kindle and from the top of the home screen, click on "More". From there, you'll find the option "My Account". Click on it and you should find the "Register" button. However, if your initial attempt to deregister your account wasn't successful, you'll still find the "Deregister" button on this tab instead. For that, simply repeat the process of deregistering your account.

2. Once you're ready to register again, make sure you note down and memorize the email and password you're using for it. This should be easy to do as this would be the same one associated with your Amazon account.

3. After your registration is confirmed, you can then proceed to downloading and purchasing new titles to add to your collection.

DELETING KINDLE BOOKS FAQs and MORE

During the process of deleting books off of your Kindle, you might encounter unfamiliar issues or other actions you're not sure how to do. In order to provide answers to those questions, here are a few other related information you might want to brush up on.

-   "I have two e-readers and only want to delete items off of one device. What do I do?"

For this, instead of selecting "delete from all devices" simply choose "delete from this device". Doing so would only remove the content off of the device you are currently using and will not affect the others.

Note that this does not apply to deregistering. Once you deregister your account, the content on all of your linked devices will be removed so be careful in doing so.

-   "The books I deleted on my Amazon page are still showing up in my Kindle. What do I do?"

Note that merely deleting it from Amazon's page does not automatically remove it from your device. In order for these changes to reflect, you can either sync your device so it automatically applies the deletion or opt to do it by yourself manually by following the instructions above.

-   "How do I remove my internet searches off of Kindle Fire?"

Go to your home screen and look for the "Silk" web browser app. From there, select the "Menu" button which is located on the upper-left corner of your screen. Once this opens, choose "Settings" then click on "Privacy". After that opens, click on "Clear Browsing Data" and simply choose the options you want to delete.

- "Is it possible for books to disappear from my library without having deleted them?"

There are instances wherein this does happen and it can be frustrating for anyone who experiences it. After all, if you're an avid collector of digital literature and do intend to read all of the titles you have saved—suddenly finding them gone can cause some serious stress. Thankfully, there are easy fixes for this that you can try out.

SYNC YOUR FIRE TABLET THEN REBOOT

Begin by following the steps for syncing your device. After, shut it down and give it a minute or two before switching it back on. You'll find that your books should start appearing one by one again—but it can also take a bit of a wait for that to happen. 20 minutes should be more than enough time to wait for all of your books to be fully functional again.

UPDATE FIRE OS VERSION

This is one of the most commonly used methods when it comes to missing Kindle books. It is also one of the easiest to do, for as long as you have a stable internet connection. To get started:

- Swipe down from the top of your home screen and choose "Settings". This should open a new menu.

- From there, scroll down and click on "Device Options".

- After, tap on "System updates" then click "Check Now / Update".

Note that there will be a bit of a wait in order for the updates to finish, but your missing books should be restored shortly after.

FACTORY RESET

Consider this a last ditch attempt when it comes to recovering missing books from your Kindle. This should only be done if the previous methods did not work for you or if you have spoken with an Amazon customer service representative and have been advised to give this a try.

What's the catch? Well, you have to understand that a factory reset will effectively remove any content you have downloaded to your tablet. This includes personal content such as your passwords, 3rd party app datas, and any side loaded content. Needless to say, you will be downloading all of those things along with any books that you want back. Sounds tedious, but if your e-reader is being problematic then this could be the solution you're looking for.

To get started:

-   Swipe down from the top of your home screen and click on "Settings".

-   Doing this should open a new menu and from it, choose "Device Options".

-   Next, click on "Reset to Factory Defaults" then click on "Reset to confirm".

Remember that there's no turning back from this so make sure you save any passwords you want to keep. It would also be helpful to create a list of the apps and books you want to re-download after the factory reset is completely. This helps ensure that you can restore your Kindle to how it was prior to resetting.

KEEPING YOUR KINDLE COLLECTION ORGANIZED

Once you have managed to delete all of the expired or read books in your collection, consider sprucing things up and organizing everything as well. This should enable you to better see what titles you have left and categorize them according to different things. For example, you can create a group for books you want to read per month or create one for books that you simply want to skim.

By doing this, you reduce the amount of frustration and stress associated with all these unread titles. After all, some of us have managed to accumulate hundreds to even thousands of books on Kindle—and whilst there's nothing wrong with wanting to read a lot, some form of organization is key to making things more accessible for ourselves.

Luckily, we can go through and organize our books through different devices—whether that be on your e-reader, your phone, your tablet, or your desktop. You can use whichever device is most comfortable for you.

KINDLE COLLECTIONS

Now, personally, I always make sure I create folders for all of my books. These keeps me from forgetting titles—which often leads to wasted money if I only end up deleting them in the end. By having this arrangement, I can also plan my reading schedules and ensure that I have a good amount of books to read each month.

Of course, our taste for literature tends to vary as well and as such, you would want different categories for the genres you read the most. To help you better visualize this set-up, consider creating digital book collections according to the following:

- Books I have finished reading

- Books I plan on reading for the month of _____

- Books that I am interested in

- Books for children

- Books I plan on lending or recommending

- Books my favorite author

- Books by ____

- Best-Sellers

- Fiction

- Non-fiction

- Reference books

- Magazines

- Zines

Categorizing your collection, especially if you have a rather huge one, can take up a bit of time so it's totally fine to do this bit by bit. You can start by scrolling through your titles and write down the different categories you find before sorting them one by one.

When it comes to creating categories, you might want to limit yourself to around 15 or so very specific ones. Within those categories, you can always create folders to sort the titles to even more specific sections. Again, this might sound tedious, but you will be thanking yourself once you've managed to organize everything. Not only will it be easier for you to find the books you need, you'll also reduce the incidents of purchasing books that are too similar to what you already have.

It pays to know what's in your library, of course.

Now, you need not worry about doing this on EVERY device you own. Because collections will sync to your Cloud account, the collections you create on one device would automatically be applied to others. Another great thing about this option is that you can add the same book to different collections—so if it falls under more than one category, you won't get confused as to where to put it or where to look for it. Convenient, right?

Worried about accidentally deleting books? Don't fret. If you do accidentally slip up and delete an entire collection, the books added to it will not be deleted.

CREATING KINDLE COLLECTIONS

There are different methods for doing this, so opt for one that best suits your preference.

METHOD 1:

- Open your Amazon account and click on "Manage Your Content and Devices".

- After the menu opens, choose "Your Content" which should then be followed by "Collections".

- This will open a drop down list where you'll find the "Create New Collection" button.

- Click this and a pop up should appear, prompting you to name the collection. Add the label then click on OK.

- And that's it, you're done!

- Once you have your collection done, all that's left to do is add books to it. For this, simply click the button placed before the title then choose the "Add to Collection" option.

- A pop up will appear, letting you choose between adding it to an existing collection or if you want to add it to a new one.

- Select the collection or collections you want to add it to then click OK.

- Once you're done, all you need to do is sync up your devices to make sure all the changes you have made gets applied to every device associated with the account.

NOTE: Amazon will only allow you to manage the books which you purchased through their store.

METHOD 2:

Aside from creating collections on the Amazon website using your desktop, you can also do it manually using your Kindle device. The process is just as straightforward. Here's what you need to do:

- Start by opening your home screen. From there, select "Create New Collection" from the menu. Much like the steps above, you have to input the label you want for the collection. You can include other details such as specific author names, dates, genres, and so on.

- Next, you can add your books to the collection. For this, simply press and hold the eBook over. A menu should pop up and from there, select "Add to Collection". Choose the collection you want to add the book to, then press "Done" at the bottom of the screen.

Remember that the content of your collections isn't permanent—you can always add or remove titles as you go. Note that any deleted books will automatically be removed from the collection it was in as well.

PRO TIP: You might notice that some books appear in "Documents from My Items" but not in "Books from My Items". The documents are the digital books you send to your Kindle device through email that were not purchased directly off of Amazon. You will not be able to organize these books into your collections like the others, unfortunately.

METHOD 3:

This one is for creating collections on Kindle using the PC app.

- Start by logging into your account. Next, right click on your chosen book's cover then select the option "Add or Remove from Collection" once the drop down list pops up.

- Next, click on "New Collection" and add the label you want for it. If you've made collections prior, you can skip adding the label and simply add it to an existing one.

This particular method is great for people who have a lot of books in their library—the type where doing it manually would be very tedious and time consuming.

Another option you can do is importing your created collections to the Kindle Paperwhite (Kindle Touch). Here's what you need to do:

- Again, login to your account—make sure it's the same account you're using for the other devices. Otherwise, it will not sync at all.

- Once you've logged in, click on "Home" then open the menu. Choose "Sync and Check for Items" from the different options.

- Next, switch to Cloud View which should allow you to open another menu. From there, choose "Import Collections". Click on "OK" on the pop up that will appear. The process should take a minute or two, but you should soon see your books and collections appearing in order.

Should you encounter issues with this process, simply reboot your device and try again.

METHOD 4:

Still find the previously mentioned methods too tedious for trying to organize your immense digital library? Well, fret not for there is a very useful add-on that you can use for this purpose. Kindlian works on Windows XP/Vista7 and 8. It also supports Kindle Keyboard, Kindle Grey, Kindle Black, Kindle Paperwhite and Kindle Touch.

This is a program that not only allows you to important and manage your Kindle collections and books, it will let you do so in a more efficient manner. Here's how:

- Choose the Kindle version you're currently using. After, a prompt should appear asking you to disconnect Kindle.

- Click on "Settings" from the Kindle menu, then choose "Update your Kindle". Reconnect your device to the computer then press OK

- Doing the previous step should make your device go into scan mode. Wait until the new interface appears.

- Once the new interface appears, you can start creating a new collection, click on "+" and add labels to what you're making. Once you've finished each one, you can simply drag the books into every collection folder you have created.

- Have epub books that you want to add to your device? Kindlian will help take care of that for you as well. Simply add them by clicking on "Add Files to Kindle" and each book will be automatically converted to the right file type.

- After you've finished with categorizing each book and adding new ones from epub, click on the "Disconnect" icon and click OK on the prompt to confirm the action.

- Lastly, after disconnecting your Kindle from your computer, click on "Settings" then "Update Your Kindle". Next, click on "Restart" and you're done.

This method works with non-Amazon books, unlike the previous ones. So if you have plenty of non-Amazon books in your collection as well, you might want to consider giving this method a try!

Conclusion

I'd like to thank you and congratulate you for transiting my lines from start to finish.

I hope this book was able to help you to learn how to delete books off of Kindle and how to keep your reader properly organized as well.

The next step is to apply what you have learned and organize your account so you can enjoy it more.

I wish you the best of luck!

To your success,

William Seals

www.ingramcontent.com/pod-product-compliance
Lightning Source LLC
Chambersburg PA
CBHW031508210526
45463CB00003B/1141